Bond

Maths
Assessment Papers

7–8 years

**J M Bond and
Andrew Baines**

Nelson Thornes

First published in 1973 by:
Thomas Nelson and Sons Ltd

This edition published in 2007 by:
Nelson Thornes Ltd
Delta Place
27 Bath Road
CHELTENHAM
GL53 7TH
United Kingdom

12 13 / 10 9 8 7 6 5 4 3 2

A catalogue record for this book is available from the British Library

ISBN 978 1 4085 1581 5

Page make-up by Tech Set Ltd

Printed and bound in Egypt by Sahara Printing Company

Before you get started

What is Bond?

This book is part of the Bond Assessment Papers series for maths, which provides **thorough and continuous practice of all the key maths content** from ages five to thirteen. Bond's maths resources are ideal preparation for many different kinds of tests and exams – from SATs to 11+ and other secondary school selection exams.

What does this book cover?

It covers all the maths that a child of this age would be expected to learn and is fully in line with the National Curriculum for maths and the National Numeracy Strategy. One of the key features of Bond Assessment Papers is that each one practises **a wide variety of skills and question types** so that children are always challenged to think – and don't get bored repeating the same question type again and again. We think that variety is the key to effective learning. It helps children 'think on their feet' and cope with the unexpected.

The age given on the cover is for guidance only. As the papers are designed to be reasonably challenging for the age group, any one child may naturally find him or herself working above or below the stated age. The important thing is that childen are always encouraged by their performance. Working at the right level is the key to this.

What does the book contain?

- **22 papers** – each one contains 30 questions.
- **Scoring devices** – there are score boxes in the margins and a Progress Chart on page 48. The chart is a visual and motivating way for children to see how they are doing. Encouraging a child to colour in the chart as they go along and to try to beat their last score can be highly effective!
- **Next Steps** – advice on what to do after finishing the papers can be found on the inside back cover.
- **Answers** – located in an easily-removed central pull-out section.
- **Key maths words** – on page 1 you will find a glossary of special key words that are used in the papers. These are highlighted in bold each time that they appear. These words are now used in the maths curriculum and children are expected to know them at this age.

How can you use this book?

One of the great strengths of Bond Assessment Papers is their flexibility. They can be used at home, school and by tutors to:

- provide regular maths practice in **bite-sized chunks**
- **highlight strengths and weaknesses** in the core skills

- identify **individual needs**
- set **homework**
- set **timed formal practice** tests – allow about 30 minutes.

It is best to start at the beginning and work though the papers in order.

What does a score mean and how can it be improved?

If children colour in the Progress Chart on page 48, this will give you an idea of how they are doing. The Next Steps inside the back cover will help you to decide what to do next to help a child progress. We suggest that it is always valuable to go over any wrong answers with children.

Don't forget the website…!

Visit www.bond11plus.co.uk for lots of advice, information and suggestions on everything to do with Bond, exams, and helping children to do their best.

Key words

Some special maths words are used in this book. You will find them in **bold** each time they appear in the papers. These words are explained here.

bar chart	a chart that records information in bars
capacity	how much liquid a container will hold. Capacity is usually measured in litres and millilitres
digit	any single number, e.g. 4 has one digit, 37 has two, 437 has three
even numbers	numbers that can be divided by two: 2, 4, 6, 8 are even numbers
fraction	a part of a whole, written like this: $\frac{1}{2}, \frac{1}{4}, \frac{2}{3}$ $\frac{2}{3}$ means two parts out of three.
frequency table	a table that records the number of times something happens
lowest term	the simplest you can make a fraction, e.g. $\frac{4}{10}$ reduced to the lowest term is $\frac{2}{5}$
mirror line	the line in which a shape can be reflected, like the reflection in a mirror
multiple	a number which another number multiplies into: 3, 6, 9, 12, 15, 60, 93 are multiples of 3
number track	a continuous strip of numbers for counting along
odd numbers	numbers that cannot be divided by two: 1, 3, 5, 7, 9 are odd numbers
pictogram	a diagram that records something using pictures
prism	a shape which has the same section all the way through, e.g. a 'tent' shape is a triangular prism
product	the answer when you multiply two numbers together: the product of 4 and 2 is 8
quadrilateral	any shape with four straight sides
rectangle, rectangular	a quadrilateral with square corners, usually with two short sides and two long sides
right angle	an angle that is equal to one quarter of a whole turn
round down, round up	round means roughly or approximately. 42 rounded down to the nearest 10 is 40. 47 rounded up to the nearest 10 is 50
standard units of measure	e.g. kilogram (kg), gram (g), metre (m), millimetre (mm) The following are not standard units of measure: cupful, handful
sum	the answer when you add two or more numbers together: the sum of 2 and 4 is 6
symmetry	a shape has symmetry if it has one or more mirror lines
Venn diagram	a chart for sorting information of different kinds
vertex, vertices	the corner or point of a shape, where its sides or faces meet. The arrows are pointing to the vertices on these two shapes

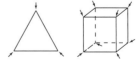

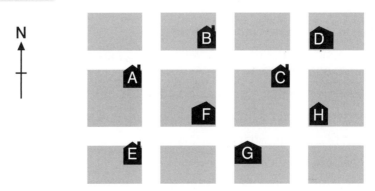

Answer the following using either: North (N), South (S), East (E) or West (W).

1 House C is ____ E ____ of house A.

2 Shop F is ____ W ____ of newsagents H.

3 School G is __ W or E __ and South of house A.

4-5 Shop D is __ SW __ and __ SW __ of house E.

6-7 House C is ____ W ____ and __ SE __ of house B.

8-9 House E is ____ S ____ and __ SE __ of house B.

10-11 House B is ____ E ____ and ____ S ____ of newsagents H.

Class 3B made a weather chart for the first 4 weeks of the term.

They drew if it was sunny, if it rained and if it was dull.

12 In the first week how many days did it rain? ___2___

13 In the second week how many days were sunny? ___3___

14 In the third week how many days were dull? ___2___

15 In the fourth week how many days did it not rain? ___6___

16 How many days were sunny in total? _14_

17 How many days were rainy in total? _6_

18 How many days were dull in total? _8_ ⑦

Draw in the missing shape on each of the following patterns.

19

20

21

22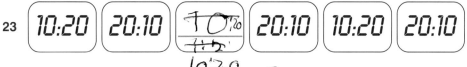

23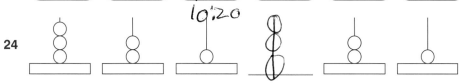

24 ⑥

How many minutes are there between:

25 10:00 and 10:30 _30_ minutes

26 10:40 and 10:55 _2_ minutes

27 11:10 and 11:30 _4_ minutes ③

28 53 subtract 30 equals _23_

29 100 less than 452 is _352_

30 Which is more: 132 cm or 123 cm? _123cm_ ③

Now go to the Progress Chart to record your score!   Total ⃝ 30

Paper 2

Here is a **frequency table** showing a class's favourite wild animals.

Favourite animal	Votes
Lion	6
Elephant	9
Giraffe	5
Zebra	3
Tiger	4

1 How many children voted for lion?

2 How many children voted for elephant?

3 How many children voted for zebra?

4 How many children voted for giraffe?

5 How many children voted for tiger?

6 How many more children voted for elephant than zebra?

7 How many more children voted for lion than giraffe?

6 ✓
9 ✓
3 ✓
5 ✓
4 ✓
6 ✓
1 ✓   7

This sign < means less than and this sign > means greater than.

Put one of these signs in each of the spaces below.

8 23 $<$ 24

9 17 $>$ 16

10 21 $>$ 19

11 10 $>$ 9

12 12 $<$ 14

13 18 $<$ 81

6

Write the answers to these questions.

14
$$\begin{array}{r} 14 \\ + 17 \\ \hline 31 \end{array}$$
A: 31

15
$$\begin{array}{r} 12 \\ + 27 \\ \hline 39 \end{array}$$
A: 39

16
$$\begin{array}{r} 15 \\ + 15 \\ \hline 30 \end{array}$$
A: 30

17
$$\begin{array}{r} 37 \\ - 14 \\ \hline 51 \end{array}$$
A: 51

18   47   A: 69
     − 22
     ___
     69

19   62   A: 94
     − 32
     ___
     94

20–22 Write down what numbers are coming out of the factory.

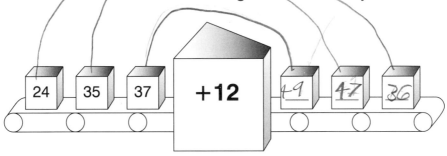

| 24 | 35 | 37 | **+12** | 49 | 47 | 36 |

23 How many times can I take 6 from 18?   3

24 I share 16 apples among 4 children.
How many apples will each child have?   4

Look at this diagram.

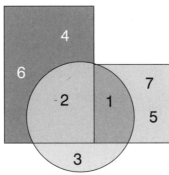

25 What is the **sum** of all the numbers in the square?   13

26 What is the **sum** of all the numbers in the circle?   6

27 What is the **sum** of all the numbers in the rectangle?   12

28 Which number is in the square and the circle?   1

29–30 Ring the numbers which are **multiples** of 10.

63   4   (30)   26   (40)

Now go to the Progress Chart to record your score!   Total   30

5

Paper 3

1–10 Complete the table below.

Multiples of 3 up to 31	
Even	**Odd**
6	3
12	9
18	15
24	21
30	27
36	33

11 How many **multiples** of 12 are there in your table?

3

12 What is 57 to the nearest 10?

60

13 What is 85 to the nearest 10?

90

14 What is 12 to the nearest 10?

10

15 What is 567 to the nearest 100?

600

16 Write the correct number in the box.

67 $\longrightarrow$ 10 more is 77

17 Write the correct number in the box.

99 $\longrightarrow$ 10 less is 89

18 Write the correct number in the box.

89 $\longrightarrow$ 20 less is 69

Write the time the clocks show.

19

2 O'clock

20

5 O'clock

21

half PAST 1

22

half PAST 10

23

12 o'clok

24

10 O'clock

6

25 17 + 6 = 23

26 37 + 6 = 43

27 67 + 6 = 73

28 87 + 6 = 93

4

29 How many 6s are there in 24?

30 How many 7s are there in 63?

$$\frac{4}{8}$$

2

Now go to the Progress Chart to record your score!   Total   **30**

7

Paper 4

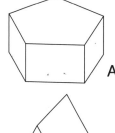 A

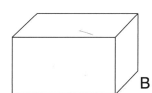 B

 C

 D

1 Which shape is not a **prism**? _____ D ✓

2–3 Which ~~two~~ *Three* shapes have more than one **vertex**? _C_, _B_ and _A_ ✓

4 Which shape has only 3 **rectangular** ~~sides~~ *faces*? _____ C ✓

5 Name shape D. _____ Cone ✓

 5

How many tens are there in each of these numbers?

6 90 = _9_ tens ✓

7 180 = _18_ tens ✓

8 160 = _16_ tens ✓

9 130 = _13_ tens ✓

10 210 = _21_ tens ✓

5

Write the answers to these questions.

11
$$\begin{array}{r} 8 \\ \times\ 5 \\ \hline 40 \end{array}$$ ✓

12
$$\begin{array}{r} 7 \\ \times\ 5 \\ \hline 35 \end{array}$$ ✓

13
$$\begin{array}{r} 6 \\ \times\ 6 \\ \hline 36 \end{array}$$ ✓

14
$$\begin{array}{r} 5 \\ \times\ 4 \\ \hline 20 \end{array}$$ ✓

4

15–16 Ring the **odd numbers**.

46 (49) (55) 58 62 74 80

17–18 Ring the **even numbers**.

(46) 49 55 (58) 61 73 81

Not to scale

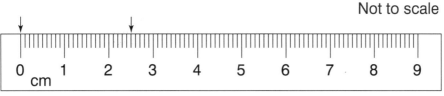

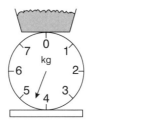

19 What is the length shown on the ruler? 2½ cm

20 What is the weight shown on the scales? 4½ kg

21 What is the **capacity** of the jug? 1000 ml

Suggest the best **standard unit of measure** for the following by choosing from km, kg, l or m:

22 how much water is in a bath   L

23 how heavy a car is   kg

24 distance from New York to Paris   Km

25 the length of a football pitch   M

26–30 Work out how much was in each person's change box.

Name	Number of 5p coins	Number of 2p coins	Number of 1p coins	Total in the change box
Caitlin	4	3	2	28p
Melly	2	4	3	21p
Richard	6	1	2	34p
Terry	5	3	1	32p
Karen	3	3	3	24p

Paper 5

Here is a **bar chart** showing the favourite colours of a group of children.

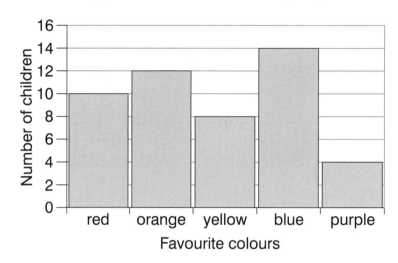

1 What was the least popular colour?   _purple_ ✓

2 What was the third most popular colour?   _red_ ✓

3 How many children voted for purple?   _4_ ✓

4 How many children voted for yellow and orange?   Y=8 O=12 ✓ (4) 4

5 What is 138 to the nearest 100?   _100_

6 What is 394 to the nearest 100?   _400_

7 The TV programme lasted 28 minutes, which is _30_ min to the nearest 10 min.

8 The computer keyboard is 33 cm long, which is _30_ cm to the nearest 10 cm.   (4) 4

9–13 Write the missing number in each space below.

36	33	30	_27_	24	21	18
74	73	72	71	70	_69_	68
97	87	77	67	_57_	47	37
24	26	_28_	30	32	34	36
26	24	22	20	_18_	16	14	(4) 5

_10_

14–19 Here is the plan of a classroom. It has been divided into squares. Show where each child sits by writing the first letter of each child's name in the correct square on the plan.

Teresa sits in D6.

Alan sits in B3.

Carol sits in E1.

Rafiq sits in A5.

Lulu sits in F4.

Giles sits in C2.

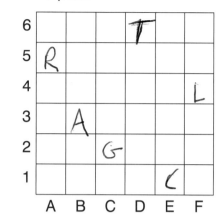

Look at this example.

$$\blacklozenge = 8 \div 4$$

$$\blacklozenge = 2$$

Now do these.

20 $\clubsuit = 5 + 6$ ↯↯

$\clubsuit = \underline{11}$

21 $\heartsuit = 3 \times 3$

$\heartsuit = \underline{9}$

22 $\spadesuit = 1 + 2 + 3$

$\spadesuit = \underline{6}$

23 $\bullet = 7 - 1$

$\bullet = \underline{6}$

24 $\Phi = 10 - 2 - 1$

$\Phi = \underline{7}$

25 $\clubsuit = 8 \div 2$

$\clubsuit = \underline{4}$

Underline the correct answer in each line.

26 $18 + 8 =$		16	28	<u>26</u>	36
27 $23 - 8 =$		31	<u>15</u>	13	14
28 £1.00 − 20p =		120p	60p	70p	<u>80p</u>
29 $\frac{1}{2}$ of 8 =		<u>4</u>	16	3	28
30 $3 + 4 + 5 =$		11	<u>12</u>	13	10

Now go to the Progress Chart to record your score! Total 29 | 30

11

Paper 6

Broken arm

This sign $<$ means less than and this sign $>$ means greater than.

Put one of these signs in each of the spaces below.

1 11 $>$ 10

2 17 $<$ 23

3 6 $<$ 60

4 8 $>$ (2 × 3)

5 7 $<$ (10 − 1)

5

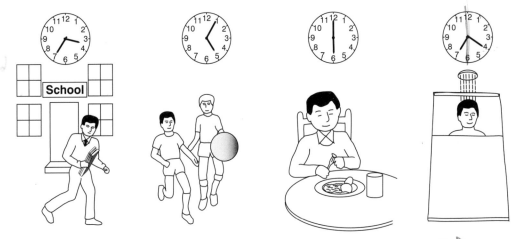

6 Michael is leaving school. What time is it? **3** : **35** p.m.

7 Michael is going out to play. What time is it? **5** : **05** p.m.

8 Michael is eating his dinner. What time is it? **6** : **0 0** p.m.

9 At dinner time, how long is it since Michael left school? **2** hr **25** min

10 Michael is having a shower. What time is it? **7** : **2 0** p.m.

5

11–16 Which numbers are coming out of these factories?

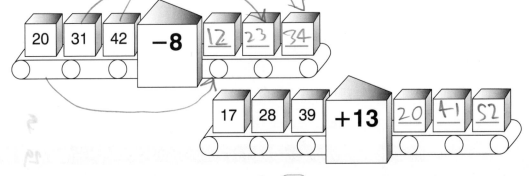

6

The following numbers are written in words.

Write them in figures.

17 One hundred and one   101

18 Two hundred and twenty   220

19 Eighty-nine   89

20 Four hundred and forty-four   444

21 Three hundred and eleven   311

22 One hundred and ten   110   ( ) 6

23–24 Draw the lines of **symmetry** (**mirror lines**) in these shapes.

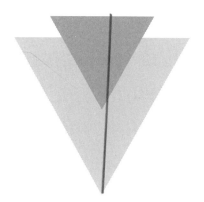 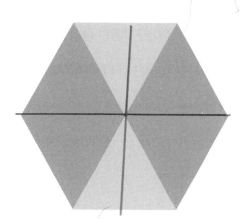

( ) 2

25 What is the biggest number you can make with these **digits:** 4 3 2
2, 3, 4?

26 Now write it in words.   Four hundred and thirty-two

27 What is the smallest number you can make with these **digits:** 3 4 5
3, 5, 4?

28 Now write it in words.   three hundred and sixty-four

29 What is the smallest number you can make with these **digits:** 5 8 9
9, 5, 8?

30 Now write it in words.   five hundred and eighty-nine   ( ) 6

Paper 7

Boys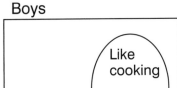

Girls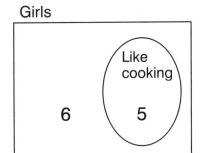

Like cooking

7   5

Like cooking

6   5

1 How many boys like cooking? _____

2 How many girls like cooking? _____

3 How many boys don't like cooking? _____

4 How many girls don't like cooking? _____

5 How many boys and girls like cooking? _____

6 How many boys and girls don't like cooking? _____ 6

Find a pair of numbers with:

7 a **sum** of 5 and a **product** of 4.   __1__ and _____

8–9 a **sum** of 5 and a **product** of 6.   _____ and _____

10–11 a **sum** of 11 and a **product** of 30.   _____ and _____

12–13 a **sum** of 19 and a **product** of 34.   _____ and _____ 7

Underline the correct answer in each line.

14 $5 + 2 + 6 =$ 11   12   13   14

15 $6 \times 10 =$ 16   60   600   66 2

14

16 Maria had 7 sweets, Kim had 3 sweets and Sara had 8 sweets.

They put them all together and then shared them equally.

How many sweets does each girl have now? _____

17 I have 39p. How much more do I need to buy a book
costing 50p? _____ p

18 Bill has 14p and Bob has 23p.

How much more has Bob than Bill? _____ p

19 Write two hundred and two in figures. _____

20 Justine has 1p more than Simon, who has 5p.

How much money do they have altogether? _____ p **5**

Last Tuesday Jason got up at 8 a.m. and went to bed at 8 p.m.

Here is a **bar chart** which shows how he spent the day.

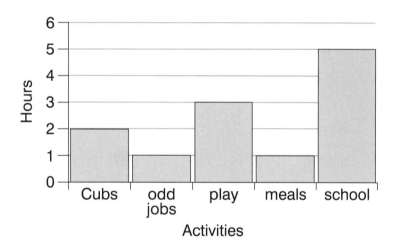

21 How long was Jason at school? _____

22 How long did he play? _____

23 How long did he spend eating? _____

24 How long did he spend at Cubs? _____

25 How long was he doing odd jobs? _____

26 How many hours did he spend out of bed? _____ **6**

27	20 − 13 ────	28	30 − 14 ────
29	20 − 11 ────	30	30 − 16 ────

Now go to the Progress Chart to record your score!   Total 30

4

Paper 8

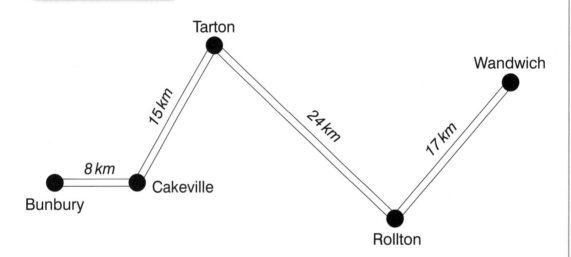

1 How far is it from Bunbury to Tarton? _____ km

2 From Tarton to Wandwich is _____ km

3 Cakeville is _____ km closer to Bunbury than it is to Tarton.

4 Rollton is _____ km closer to Wandwich than it is to Tarton.

5 How far is it from Bunbury to Rollton? _____ km

6 How far is it from Wandwich to Cakeville? _____ km

7 How far is it from Rollton to Cakeville? _____ km

8–9 Which two towns are nearest to each other? _____ and _____

9

10 How many 6s are there in 120? _____

11 How much smaller is 27 than 80? _____

12 How much have I altogether if I have £1.00, 25p and £3.10? _____

13 If I make 75 three times as big, what number will it be? _____

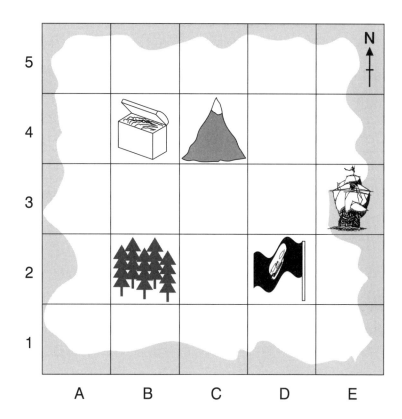

There is a flag on the island at D2.

14 Where has the pirate ship landed? _____

15 Where is the forest? _____

16 Where is the mountain? _____

17 Where is the treasure? _____

18 Mark a big X in the middle of C3.

Use either North (N), South (S), East (E) or West (W) to complete the following sentences.

19 The treasure is _____ of the mountain.

20 The treasure is _____ of the forest.

21 The flag is _____ of the forest.

Write the missing number in each space below.

22	4	8	12	_____	20
23	15	_____	9	6	3
24	5	10	15	_____	25
25	24	_____	16	12	8

4

26–30 A train takes 1 hr 5 min to travel from Coldville to Warmwich.

Fill in this chart.

| Leaves Coldville | 06:00 | 07:05 | 08:15 | 09:35 | 10:55 |
| Arrives Warmwich | | | | | |

5

Now go to the Progress Chart to record your score!   Total 30

Paper 9

1–6 Put the letters **a** to **f** in the correct place in the **Venn diagram**.

Shapes

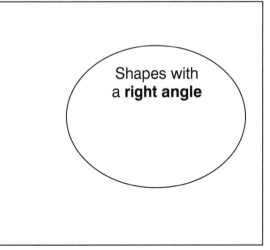

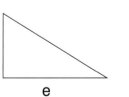

7 Shape **c** is called a _____

7

Write the missing number in each space below.

| 8 | 6 | 9 | 12 | 15 | _____ |

| 9 | 80 | 90 | _____ | 110 | 120 |

2

Write the answers to these calculations.

10   £ 1.20
    + £ 2.35
    _____

11   £ 1.05
    + £ 2.06
    _____

12   £ 2.12
    + £ 1.21
    _____

13   £ 3.02
    + £ 1.08
    _____

4

14–15 Here is part of a **number track**. Write 254 and 243 on it.

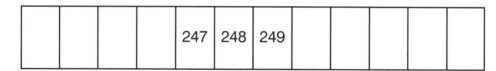

| | | | | 247 | 248 | 249 | | | | | | |

16–17 Here is part of a **number track**. Write 583 and 592 on it.

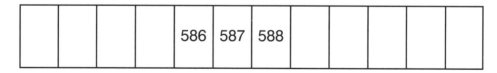

| | | | | 586 | 587 | 588 | | | | | | |

4

18–19 Ring the **odd numbers**.

26   79   35   58   92   14   60

20–21 Ring the **even numbers**.

25   59   15   38   61   73   90

4

Here is a **bar chart** some boys made.

It shows their heights.

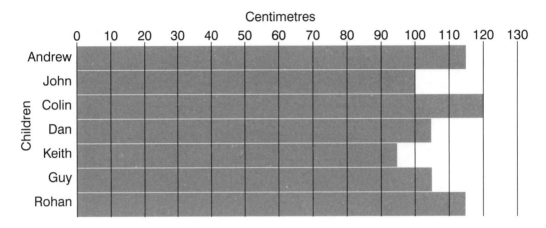

22–23 Name two boys who are the same height and are 115 cm tall.

_____ and _____

24–25 Name two boys who are 105 cm tall. _____ and _____

26 Who is the tallest boy?   _____

27 How tall is he?   _____

28 Who is the shortest boy?   _____

29 Keith is _____ cm shorter than Colin.

30 How many boys are less than 110 cm tall?   _____

Now go to the Progress Chart to record your score!   Total   30

9

Paper 10

| January is the 1st month | July is the 7th month |
| March is the 3rd month | September is the 9th month |

January is the 1st month   July is the 7th month
February is the 2nd month   August is the 8th month
March is the 3rd month   September is the 9th month
April is the 4th month   October is the 10th month
May is the 5th month   November is the 11th month
June is the 6th month   December is the 12th month

1st of November 1993 can be written as 1.11.93.

Here are some birthdays written this way.

Anna was born	20.10.95
Carl was born	21.2.92
Karen was born	23.8.94
Martin was born	24.5.93

In which month was each child born?

1 Anna   _____

2 Carl   _____

3 Karen   _____

4 Martin   _____

5 _____ is the eldest.

6 _____ is the youngest.

○ 6

Here is an easy way to add up three numbers which come after each other.

$1 + 2 + 3 = 6$ which is the same as $\boxed{3 \times 2}$ = 6

$5 + 6 + 7 = 18$ which is the same as $\boxed{3 \times 6}$ = 18

Now do these in the same way.

7 $2 + 3 + 4$ [   ] = _____

8 $9 + 10 + 11$ [   ] = _____

9 $4 + 5 + 6$ [   ] = _____

10 $6 + 7 + 8$ [   ] = _____

11 $12 + 11 + 10$ [   ] = _____

12 $7 + 8 + 9$ [   ] = _____

○ 6

13 If Jack was 4 years older he would be the same age as Jill.

Jill is 13. How old is Jack?   _____ years

14 Add together 4 and 7 and then take 5 from your answer.

How many do you have now?   _____

○ 2

Can you complete these bills for the baker? Write your answers in £.

15–18 Mrs Smith bought:

	£
2 mince pies	_____
1 cream sponge	_____
1 cream bun	_____
Total	_____

19–22 Mrs Jones bought:

	£
2 cream buns	_____
1 fruit cake	_____
1 mince pie	_____
Total	_____

23–27 Mr Green bought:

	£
1 fruit cake	_____
1 cream sponge	_____
1 mince pie	_____
1 cream bun	_____
Total	_____

13

Measure the lengths of these items.

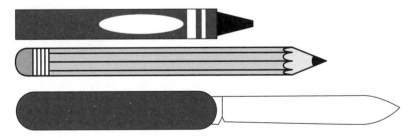

28 The crayon is _____ cm long.

29 The pencil is _____ cm long.

30 The penknife is _____ cm long.

3

Now go to the Progress Chart to record your score!   Total   30

22

Paper 1

1 East (E)
2 West (W)
3 East (E)
4–5 North (N), East (E)
6–7 South (S), East (E)
8–9 South (S), West (W)
10–11 North (N), West (W)
12 2
13 3
14 2
15 6
16 14
17 6
18 8
19 ▢
20 (50p)
21 ◯
22 ⬡
23 10:20
24 (abacus with 3 beads)
25 30
26 15
27 20
28 23
29 352
30 132 cm

Paper 2

1 6
2 9
3 3
4 5
5 4
6 6
7 1
8 <
9 >
10 >
11 >
12 <
13 <
14 31
15 39
16 30
17 23
18 25
19 30
20 36
21 47
22 49
23 3
24 4
25 13
26 6
27 12
28 1
29–30 30, 40

Paper 3

1–10

Multiples of 3 up to 31	
Even	Odd
6	3
12	9
18	15
24	21
30	27

11 2
12 60
13 90
14 10
15 600
16 77
17 99
18 69
19 2 o'clock
20 5 o'clock
21 half past 1 or One thirty
22 half past 10 or Ten thirty
23 11 o'clock
24 10 o'clock
25 23
26 43
27 73
28 93
29 4
30 9

Paper 4

1 D
2–3 B, C
4 C
5 cone
6 9
7 18
8 16
9 13
10 21
11 40
12 35
13 36
14 20
15–16 49, 55
17–18 46, 58
19 2.5
20 4.5
21 1000
22 l
23 kg
24 km
25 m
26 28p
27 21p
28 34p
29 32p
30 24p

Paper 5

1 purple
2 red
3 4
4 20
5 100
6 400
7 30
8 30
9 27
10 69
11 57
12 28
13 18
14–19

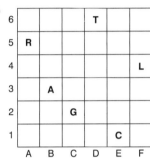

20 11
21 9
22 6
23 6
24 7
25 4
26 26
27 15
28 80p
29 4
30 12

Paper 6

1 >
2 <
3 <
4 >
5 <
6 3:35
7 5:05
8 6:00
9 2 hr 25 min
10 7:20
11 12
12 23
13 34
14 30
15 41
16 52
17 101
18 220
19 89
20 444
21 311
22 110
23

24

25 432
26 Four hundred and thirty-two
27 345
28 Three hundred and forty-five
29 589
30 Five hundred and eighty-nine

1 5
2 5
3 7
4 6
5 10
6 13
7 4
8–9 2, 3
10–11 5, 6
12–13 2, 17
14 13
15 60
16 6
17 11
18 9
19 202
20 11
21 5 hours
22 3 hours
23 1 hour
24 2 hours
25 1 hour
26 12 hours
27 7
28 16
29 9
30 14

1 23
2 41
3 7
4 7
5 47
6 56
7 39
8 Bunbury
9 Cakeville
10 20
11 53
12 £4.35
13 225
14 E3
15 B2
16 C4
17 B4
18

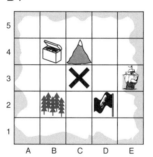

19 West (W)
20 North (N)
21 East (E)
22 16
23 12
24 20
25 20
26–30

Leaves Coldville					
Arrives Warmwich	07:05	08:10	09:20	10:40	12:00

Paper 9

1–6 Shapes

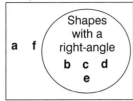

7 rectangle
8 18
9 100
10 £3.55
11 £3.11
12 £3.33
13 £4.10

14–15

243			247	248	249					254

16–17

	583		586	587	588			592	

18–19 79, 35
20–21 38, 90
22–23 Andrew, Rohan
24–25 Dan, Guy
26 Colin
27 120 cm
28 Keith
29 25
30 4 boys

Paper 10

1 October
2 February
3 August
4 May
5 Carl
6 Anna
7 9
8 30
9 15
10 21
11 33
12 24
13 9
14 6

15–18

0.50	1.35	0.35	2.20

19–22

0.70	3.50	0.25	4.45

23–27

3.50	1.35	0.25	0.35	5.45

28 6.5
29 8.5
30 10.5

Paper 11

1 55p
2 20p
3 30p
4 45p
5 25p
6 30p
7 30p
8 25p
9 500
10 200
11 40 min
12 40 cm
13 80 cm
14 36
15 15
16 300
17 40
18 60
19 5
20 75p or £0.75
21 £3.75
22 1994
23 6
24 3
25 5
26 7
27 11:15
28 10:30
29 7:45
30 12:10

Paper 12

1 500
2 2
3 800
4–5 C, D
6 5
7–8 (97), 67
9–10 (87), 46
11-12 (55), 35
13 10 + 1
14 6 − 4
15 2 + 1
16 2 × 1
17 50
18 70
19 60
20 3
21 5
22 5
23 4
24 6
25 4
26 2
27-28 3 and 6
29-30 1 and 11

Paper 13

1–7 rectangle
pyramid
cube
cone
cuboid
triangle
square

8–9 (43), 23
10–11 (77), 17
12–13 (53), 31
14 +
15 −
16 ×
17
18
19
20 2000
21 $\frac{1}{4}$ 22 $\frac{3}{4}$
23 $\frac{3}{4}$ 24 $\frac{1}{4}$
25 $\frac{1}{2}$ 26 $\frac{1}{2}$
27 $\frac{1}{2}$ 28 $\frac{1}{2}$
29 247
30 two hundred and forty-seven

Paper 14

1

80	24

3

= 104

2

60	30

5

= 90

3

6	60	48

8

=108

4

5	50	45

10   9

= 95

5

6	60	42

10   7

= 102

6

9	90	45

10   5

= 135

7 8
8 10
9 10
10 5
11 5
12 3
13 10
14 35
15 999
16 3
17 5
18 2
19 4
20 19
21 15 min
22 7.5 cm
23 11 cm
24 12 cm
25 5.5 cm
26 9 cm
27 10.5 cm
28 25
29 102
30 120

Paper 15

1 20
2 100
3 20
4 10
5 $\frac{1}{2}$
6 $\frac{1}{2}$
7 $\frac{1}{4}$
8 $\frac{3}{4}$
9 $\frac{1}{2}$
10 $\frac{1}{2}$
11 $\frac{1}{2}$
12 $\frac{1}{2}$
13 $\times$
14 $-$
15 $\times$
16 6
17 6
18 7
19 7
20 01:10
21 40
22 25
23 45
24 30
25 52
26 20
27 25
28 250
29 30
30 2.5 or $2\frac{1}{2}$

Paper 16

1  (8 o'clock)

2 (twenty to four)

3 (ten past five)

4 (quarter to eleven)

5 5
6 8
7 3
8 4
9 3
10 5

11–16

Nicholas	4	6	**9**	11	**14**	17
Annabel	**9**	**11**	14	**16**	19	**22**

12 11
13 9
14 16
15 14
16 22
17 70
18 40
19 26

20

Sunny						
Wet						
Dull						

21 3
22 5
23 2
24 21
25 44
26 58

27–30

24	28	27	30

Paper 17

1. £2.40
2. 60p or £0.60
3. crayons and eraser
4. £1.70
5. 30p or £0.30
6. £3.10
7. £1.95
8. felt-tip pen set and scissors
9. 43
10. 42
11. 66
12. 447
13. 69
14. 100
15. 456
16. 378
17. 48
18. 6
19. 12
20. 6
21. 18
22. $\frac{1}{2}$
23. $\frac{1}{4}$
24. $\frac{1}{4}$
25. 21
26. 50 min
27. 20 min
28. 30 miles
29. 300 miles
30. 600 miles

Paper 18

1. E
2. B
3. A
4. F
5. D
6. C
7. 40
8. 15
9. 35
10. 3×3
11. $10 + 1$
12. $3 + 2$
13. 3456
14. 7620
15. 1389
16. 9851
17.

	10	1	
15	150	15	= 165

18.

	10	4	
11	110	44	= 154

19.

	10	9	
11	110	99	= 209

20.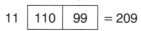
21.
22.
23. Yes
24. No
25. No
26. Yes
27. No
28. No
29. kg
30. 15

Paper 19

1. 6.5
2. 11.5
3. 13
4. 9
5. 30 min
6. 9p
7. 20
8. 17
9. 18
10. 20
11. 21
12. Shaun
13. Zanna
14. Peter
15. Tim
16. Rachel
17. 100 m
18. 250 m
19. 500 m
20. 500 m
21. 800 m
22.
23.
24. 50 min
25. 1 hr 30 min or 90 min
26. 43
27. 20
28. 19
29–30. 8 and 5

Paper 20

1 06:20
2 07:30
3 09:55
4 11:00
5 12:25
6 $\frac{3}{4}$
7 $\frac{1}{4}$
8 $\frac{1}{2}$
9 $\frac{1}{4}$
10 $\frac{1}{2}$
11 $\frac{1}{4}$
12 2
13 7
14 6
15 26
16 5 + 2
17 11 + 2
18 15 × 1
19 3 × 6
20 8 + 11
21 0.5 or $\frac{1}{2}$
22 2
23 1.5 or $1\frac{1}{2}$
24 1
25 2.5 or $2\frac{1}{2}$
26 4.5 or $4\frac{1}{2}$
27 4
28 3
29 07:45
30 03:15 or 15:15

Paper 21

1 7
2 6
3 6
4 5
5 5
6 6
7 350
8 540
9 900
10 340
11 189
12 298
13 13
14 01:15
15 16 days
16 £3.37
17 £3.88
18 £5.88
19 £7.90
20 20
21 18
22 21
23 18
24 17
25 B
26 D
27 1.5 kg or $1\frac{1}{2}$ kg
28 5.5 kg or $5\frac{1}{2}$ kg
29 4.5 kg or $4\frac{1}{2}$ kg
30 20 weeks

Paper 22

1–6 Shapes

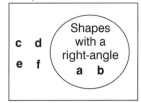

7–8 e, f
9–11 a, b, c
12 24
13 >
14 <
15 >
16 >
17   18
19 20
21 22
23 24

25 10
26 1
27 7
28 3
29

Seaside	🧍	🧍	🧍	🧍	🧍	🧍	🧍	🧍	🧍
Camp	🧍	🧍							
Farm	🧍								
Abroad	🧍	🧍	🧍	🧍	🧍	🧍			
Touring	🧍	🧍	🧍	🧍					

30 1007

Here is a chart showing how much money some children spent on fruit yesterday.

Each square represents 5p.

Sam										
Amy										
Mark										
Simon										
Jessica										
Helen										

1 How much did Sam spend?

2 How much did Amy spend?

3 Mark spent

4 Simon spent

5 Jessica spent

6 Helen spent

7 How much more did Sam spend than Jessica?

8 How much more did Simon spend than Amy? ⑧

9 What is 473 to the nearest 100?

10 What is 246 to the nearest 100?

11 The maths lesson lasted 36 minutes, which is _____ to the nearest 10 min.

12 The laptop computer is 42 cm long, which is _____ to the nearest 10 cm.

13 Two laptop computers side by side measure _____ to the nearest 10 cm. ⑤

14 Draw a line under any number that is a **multiple** of 3 but not 5.

28   35   36   25   15

15 Put a ring round any number that is a multiple of 3 and 5.

28   35   36   25   15

2

16 $30 \times 10 =$ _____

17 $4 \times 10 =$ _____

18 $600 \div 10 =$ _____

19 $50 \div 10 =$ _____

4

20 If 4 comics cost £3.00, what is the cost of one?   _____

21 How much would 5 comics cost?   _____

22 Annabel is 2 years older than Mary. Mary was born in 1996.

What year was Annabel born?   _____

3

23–26 Fill in the missing numbers to make the scales balance.

4

27–30 Write down the time shown on each clock below.

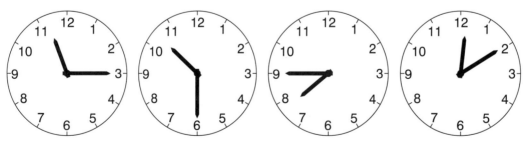

_____ : _____ a.m.   _____ : _____ p.m.   _____ : _____ a.m.   _____ : _____ p.m.

4

Now go to the Progress Chart to record your score!   Total   30

24

Paper 12

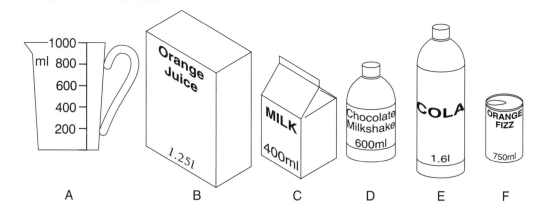

A   B   C   D   E   F

1   If the jug is half full how many ml are in it?   _____ ml

2   How many containers hold over a litre?   _____

3   If E was half full how much would it contain?   _____ ml

4–5   Which two containers when emptied would,
      together, fill the measuring jug?   _____ and _____

6   If another container holds 200 ml how many times
    could I refill it from a full jug?   _____   6

Put a ring round the highest number and draw a line under the smallest
number on each line.

7–8	79	76	77	97	67
9–10	64	87	46	68	78
11–12	55	53	54	35	45

6

One question in each line has a different answer from the others.

Draw a line under it.

13	5 × 2	10 × 1	12 − 2	10 + 1	6 + 4
14	9 − 8	1 × 1	7 − 6	10 − 9	6 − 4
15	4 ÷ 2	2 + 1	2 × 1	8 − 6	4 − 2
16	2 × 1	3 × 1	2 + 1	5 − 2	6 − 3

4

17 80 − 30 = _____

18 20 + 50 = _____

19 100 − 40 = _____

Here is part of a ruler. Look at it and then answer the questions.

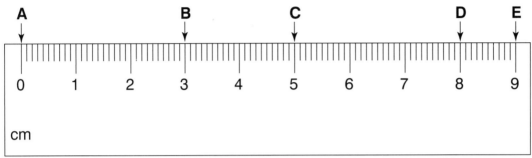

Not to scale

20 How far is it from A to B? _____ cm

21 From A to C is _____ cm.

22 From B to D is _____ cm.

23 From C to E is _____ cm.

4

Fill in the missing **digits**.

24 47 − 2_____ = 21

25 _____7 − 25 = 22

26 74 − _____3 = 51

3

Find a pair of numbers with:

27–28 a **sum** of 9 and a **product** of 18. _____ and _____

29–30 a **sum** of 12 and a **product** of 11. _____ and _____

4

Paper 13

1–7 Draw a line between each of these shapes and its name.

rectangle

pyramid

cube

cone

cuboid

triangle

square

7

Put a ring round the highest number and draw a line under the lowest number on each line.

8–9	24	23	34	32	43
10–11	19	71	77	29	17
12–13	31	35	36	53	33

6

Put a sign (+ − ÷ ×) in each space to make the answer correct.

14 $3 \_\_\_ 5 = 8$

15 $9 \_\_\_ 2 = 7$

16 $2 \_\_\_ 4 = 8$

3

Draw all the lines of **symmetry** (**mirror lines**) in these shapes.

17    **18**    **19**

3

20 Alison was 5 in 2005. In what year was she born? _____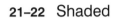

What **fraction** of each of these shapes is shaded and what **fraction** is white? Express answer in **lowest term**.

21–22 Shaded _____

  White _____

23–24 Shaded _____

  White _____

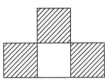

25–26 Shaded _____

  White _____

27–28 Shaded _____

  White _____

8

29 What is the lowest number you can make with these **digits**:

  4, 7, 2? _____

30 Now write it in words. _____

2

Now go to the Progress Chart to record your score!   Total 30

Paper 14

Here is an easy way to work out 14 × 8.

×	10	4
8	80	32	= 112

Now do these multiplications in the same way.

1 13 × 8

×	10	
8			= _____

2 15×6

$\times$ 10

6 | | | = _____

3 18×6

$\times$ 10

| | | = _____

4 19×5

$\times$

| | | = _____

5 17×6

$\times$

| | | = _____

6 15×9

$\times$

| | | = _____

      6

7–12 Fill in the missing numbers to make the scales balance.

| 20 − 4 | △ | 2 × |

| 30 − | △ | 4 × 5 |

| 15 + | △ | 5 × 5 |

| 10 + | △ | 5 × 3 |

| 16 ÷ 4 | △ | 9 − |

| 3 × 3 | △ | 12 − |

      6

13 Jade has 20 socks. If she puts them in pairs she will have _____ pairs.

14 How much must I add to 65p to make £1.00?   _____ p

15 What number is one less than 1000? _____ 3

There is one **digit** missing from each of these calculations.
Can you find what it is? Write it in the space.

16
$$
\begin{array}{r}
2\ \ 4 \\
+\ 1\ \underline{} \\
\hline
3\ \ 7 \\
\hline
\end{array}
$$

17
$$
\begin{array}{r}
3\ \underline{} \\
+\ 2\ \ 4 \\
\hline
5\ \ 9 \\
\hline
\end{array}
$$

18
$$
\begin{array}{r}
2\ \ 5 \\
+\ 1\ \underline{} \\
\hline
3\ \ 7 \\
\hline
\end{array}
$$

19
$$
\begin{array}{r}
\underline{}\ \ 6 \\
+\ 2\ \ 3 \\
\hline
6\ \ 9 \\
\hline
\end{array}
$$

4

20 Orla had 40p. She bought two erasers and had only 2p left. This means that one eraser costs _____ p.

21 How many minutes are there in a quarter of an hour? _____ 2

Use your ruler to measure these lines in centimetres.

22 _____

23 _____

24 _____

25 _____

26 _____

27 _____ 6

There are 51 children altogether in Classes 1 and 2.

28 If there are 26 in Class 1 how many are there in Class 2? _____

29 Write one hundred and two in figures. _____

30 Write one hundred and twenty in figures. _____ 3

Paper 15

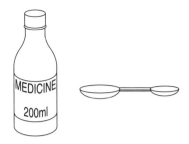

The medicine spoon has two ends. One end can measure 5 ml. The other end can measure 10 ml.

1 If the bottle is full how many doses of 10 ml are in the bottle?

_____

2 When the bottle is half full how many ml are in it? _____ ml

3 If the bottle is half full how many doses of 5 ml are in the bottle?

_____

Another much larger spoon holds 20 ml.

4 How many times can you fill this spoon from the full bottle? _____ ④ 4

What **fraction** of each of these shapes is shaded and what **fraction** is white? Express answer in **lowest term**.

5–6 Shaded _____

White _____

7–8 Shaded _____

White _____

9–10 Shaded _____

White _____

11–12 Shaded _____

White _____

⭘ 8

Put a sign $(+ - \div \times)$ in each question to make it correct.

13 4 _____ 4 = 16

14 7 _____ 2 = 5

15 3 _____ 0 = 0

Here are some questions on **rounding up** or **down**.

We know that 37 ÷ 6 = 6 remainder 1.

16 I have 37 eggs. One egg box holds 6 eggs.

I could fill only _____ egg boxes.

17 I have £37. Cinema tickets cost £6 each.

I could only buy _____ tickets.

18 I have 37 eggs. One egg box holds 6 eggs.

I will need _____ egg boxes to hold all 37 eggs.

19 I have 37 guests coming for a special dinner.

My tables can only seat 6 people each.

_____ tables are needed to seat all the guests.

20 Paul's watch is 5 minutes fast. It shows 01:15.

What is the right time?   _____ : _____

Here is a train timetable.

	Train A	Train B	Train C	Train D
Leaves Rigby	09:00	10:05	11:15	12:10
Arrives in Stairs	09:40	10:30	12:00	12:40

21 Train A takes _____ min.   **22** Train B takes _____ min.

23 Train C takes _____ min.   **24** Train D takes _____ min.

25 What number is five tens and a two?   _____

26 What number is three fours and an eight?   _____

32

27 What number is halfway between 20 and 30? _____

28 What number is halfway between 200 and 300? _____

29 What number is halfway between 25 and 35? _____

30 What number is halfway between 2 and 3? _____ 4

Now go to the Progress Chart to record your score! Total 30

Paper 16

1–4 Draw hands on each of the clocks below to show the correct time.

Make sure you put the small hand in the right place.

8 o'clock    Twenty to four

Ten past five   Quarter to eleven  4

Do you remember how to do these questions from Paper 5? Now do these.

5 $5 \times \clubsuit = 25$

$\clubsuit =$ _____

6 $2 \times \heartsuit = 16$

$\heartsuit =$ _____

7 $3 \times \spadesuit = 9$

$\spadesuit =$ _____

8 $3 \times \bullet = 10 + 2$

$\bullet =$ _____

9 $5 \times \Phi = 15$

$\Phi =$ _____

10 $2 \times \clubsuit = 8 + 2$

$\clubsuit =$ _____ 6

11–16 Annabel is 5 years older than Nicholas.

Fill in the missing ages in the chart below.

When Nicholas is	4	6		11		17
Annabel is			14		19	

6

33

Underline the correct answer in each line.

17 700 ÷ 10 =   7   70   700   7000

18 5 × 8 =   58   35   40   45

19 17 + 9 =   26   27   29   25   **3**

Last year Samir went to stay on a farm. He was away for 14 days.

He made this chart to show what the weather was like. He has not filled in the entries for dull days. Each square represents 1 day.

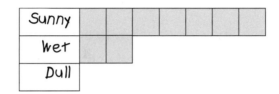

20 Draw on the chart the number of days it was dull.

21 How many more days were dull than were wet?   _____

22 How many more days were sunny than were wet?   _____

23 How many more days were sunny than were dull?   _____   **4**

Write the missing number in each line.

24   18   _____   24   27   30

25   29   34   39   _____   49

26   18   28   38   48   _____   **3**

27–30 Here are the number of hours some children spent playing football in four weeks. Find each child's total.

	Indira	Beth	Carmen	Julie
Week 1	7	8	5	9
Week 2	4	7	7	8
Week 3	6	4	8	6
Week 4	7	9	7	7
Totals				

4

Paper 17

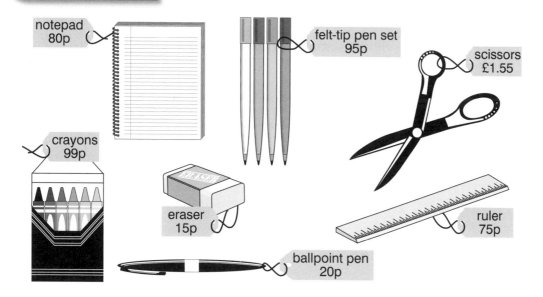

notepad 80p

felt-tip pen set 95p

scissors £1.55

crayons 99p

eraser 15p

ruler 75p

ballpoint pen 20p

1 I bought 3 notepads. What did I pay? _____

2 How much change did I get from £3.00? _____

3 Which 2 things can I buy for £1.14 exactly? _____ and _____

4 Dad bought a notepad, an eraser and a ruler.
   Together they cost _____

5 How much change did Dad get from £2.00? _____

6 How much do two pairs of scissors cost?
   (Remember one pair of scissors is one item.) _____

7 How much would it cost to buy 2 ballpoint pens
   and a pair of scissors? _____

8 Which 2 things would cost £2.50 exactly?

_____

8

9   60
   – 17
   ____

10   70
   – 28
   ____

11   80
   – 14
   ____

3

12 347 + 100 = _____

13 59 + 10 = _____

14   90 + 10 = _____

15   356 + 100 = _____

16   478 − 100 = _____

17   58 − 10 = _____

4

Here is a **pictogram** showing which summer sports children in Class 3A like best.

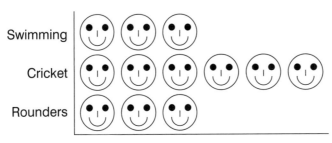

= 2 children

Swimming

Cricket

Rounders

Number of children

18   How many like swimming best? _____

19   How many like cricket best? _____

20   How many like rounders best? _____

21   How many did not choose rounders? _____

22   What **fraction** of the class likes cricket best? _____

23   What **fraction** of the class likes swimming best? _____

24   What **fraction** of the class likes rounders best? _____

7

25   What is the **sum** of 5, 7 and 9? _____

1

26   The film lasted 53 minutes, which is _____ to the nearest 10 min.

27   The music video lasted 16 minutes, which is _____ to the nearest 10 min.

28   Oxton to Muleton is 27 miles, which is _____ to the nearest 10 miles.

29   Mondon to Meeds is 283 miles, which is _____ to the nearest 100 miles.

30   The return trip (Mondon to Meeds and back) is _____ to the nearest 100 miles.

5

Paper 18

A   C E

B   D F

Put the letters A to F in the correct spaces to make the following sentences true.

1 _____ is a **prism**.

2 The sphere is _____ .

3 _____ is a circle.

4 The **quadrilateral** is _____ .

5 _____ is a semicircle.

6 The hemisphere is _____ .

6

How many minutes are there between:

7  and $\boxed{11.40}$ _____ min   **8** $\boxed{10.35}$ and  _____ min

9 and $\boxed{12.45}$ _____ min

3

One question on each line has a different answer from the others.

Put a line under it.

10	$7 + 1$	4×2	8×1	3×3	$6 + 2$
11	2×5	$11 - 1$	$10 + 1$	10×1	$4 + 6$
12	$3 + 2$	6×1	3×2	$7 - 1$	$4 + 2$

3

13 What is the smallest number you can make with these **digits**: 5, 3, 6, 4?   _____

14 What is the biggest number you can make with these **digits**: 2, 0, 6, 7?   _____

15 What is the smallest number you can make with these **digits**: 1, 9, 3, 8?   _____

16 What is the biggest number you can make with these **digits**: 5, 1, 8, 9?   _____

4

Here is an easy way to work out 16×11.

$\times$	10	1	
16	160	16	$= 176$

Now do these the same way.

17 15×11

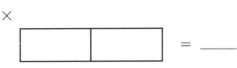

$\times$

$=$ _____

18 11×14

$\times$

$=$ _____

19 11×19

$\times$

$=$ _____

◯ 3

Draw all the lines of **symmetry** (**mirror lines**) in these shapes.

20

21

22

◯ 3

Is it possible to do the following questions? Write 'Yes' or 'No'.

23 $4\,cm + 2\,m + 17\,cm$ _____

24 $5\,m + 4\,l + 5\,l$ _____

25 $4\,km + 2\,kg + 3\,km$ _____

26 $8\,kg + 20\,g + 27\,g$ _____

27 $40\,ml + 21\,g + 5\,ml$ _____

28 $3\,cm - 11\,ml + 5\,ml$ _____

◯ 6

29 Choose the best unit of measure from the above to measure the weight of a man. _____

◯ 1

30 What is the **sum** of the numbers 1 to 5? _____

◯ 1

Measure the lengths of these things.

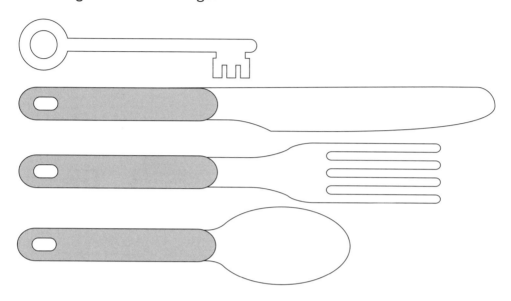

1 The key is _____ cm long.

2 The fork is _____ cm long.

3 The knife is _____ cm long.

4 The spoon is _____ cm long.   ◯ 4

5 A lesson started at ten minutes to two and finished at twenty minutes past two.

How long did the lesson last?   _____   ◯ 1

6 What must I add to 16p to make 25p?   _____   ◯ 1

Here are the marks some children scored in a test.

7–11 Add them up.

	Peter	Zanna	Tim	Rachel	Shaun
	6	7	8	6	7
	5	4	7	6	8
	9	6	3	8	6
Totals					

12 Who had the highest total mark? _____

13 Who had the lowest total mark? _____

14 Who got the highest mark in any one test? _____

15 Who got the lowest mark in any one test? _____

16 Who got the same mark in two tests? _____

10

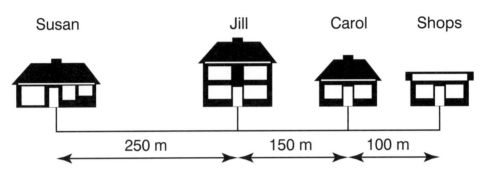

Susan   Jill   Carol   Shops

250 m   150 m   100 m

17 How far does Carol live from the shops? _____

18 How far does Jill have to walk to the shops? _____

19 How far does Susan live from the shops? _____

20 If Jill walked to Susan's house and then home again how far would she walk? _____

21 If Susan walked to Carol's house and then home again how far would she walk? _____

5

Draw all the lines of **symmetry** (**mirror lines**) in these shapes.

22

23

2

24 It takes Abdul 46 minutes to clean a car inside and out, which is _____ to the nearest 10 min.

25 How long will it take Abdul to clean two cars to the nearest 10 min? _____

2

40

26 What number is four tens and three? _____

27 What number is three sixes and two? _____

28 What number is five threes and four? _____ ◯ 3

29–30 Find a pair of numbers with a **sum** of 13 and a **product** of 40.

_____ and _____ ◯ 2

Now go to the Progress Chart to record your score! Total ◯ 30

Paper 20

1–5 A train takes 1 hr 10 min to travel from Moreton to Bidston.

Fill in the timetable.

	Train A	Train B	Train C	Train D	Train E
Leaves Moreton	05:10	06:20	08:45	09:50	11:15
Arrives at Bidston					

◯ 5

What **fraction** of each shape is shaded? Express answer in **lowest term**.

6 ___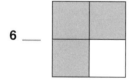

7 ___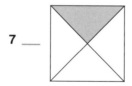

8 ___

9 ___

10 ___

11 ___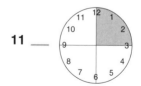

◯ 6

Find the missing **digits**.

12
```
    3 ___
  + 1   4
  ───────
    4   6
```

13
```
    1   9
  - ___
  ───────
    1   2
```

14
```
      4   2
    ×       3
  ─────────
    1   2 ___
```

◯ 3

41

15 What is half of 52? _____

Underline the incorrect answer in each line.

16	10 =	6 + 4	3 + 7	5 + 2	10 × 1
17	12 =	11 + 2	6 + 6	4 + 8	2 × 6
18	16 =	2 × 8	4 × 4	17 − 1	15 × 1
19	15 =	7 + 8	3 × 6	9 + 6	15 ÷ 1
20	18 =	8 + 11	12 + 6	3 × 6	2 × 9

This drawing shows part of a centimetre ruler. Look at it and then answer the questions.

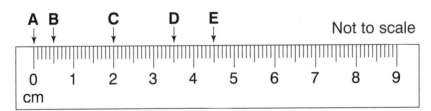

21 How far is it from A to B? _____ cm

22 From A to C is _____ cm

23 From C to D is _____ cm

24 From D to E is _____ cm

25 How far is it from C to E? _____ cm

26 From A to E is _____ cm

27 From B to E is _____ cm

28 From B to D is _____ cm

29 We will have our breakfast in a quarter of an hour.
It is now 07:30. When will we have breakfast? _____ : _____

30 It is a quarter to four. What time was it half an hour ago? _____ : _____

Now go to the Progress Chart to record your score!　Total　30

Paper 21

1–6 Fill in the spaces to make the scales balance.

 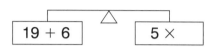

6

7 What is the **multiple** of 10 that follows 340? _____

8 What is the **multiple** of 10 that comes before 550? _____

9 What is the **multiple** of 100 that follows 800? _____

10 What is the **multiple** of 5 that comes before 345? _____

11 What **odd number** comes before 191? _____

12 What **even number** comes before 300? _____

13 What number is 4 more than 3×3? _____

14 My watch is 5 minutes slow.
If it shows 10 minutes past 1 what is the right time? _____ : _____

15 My dog eats 4 biscuits a day.
How long will a packet of 64 biscuits last him? _____

9

16 £1.24
+ £2.13
_____

17 £2.43
+ £1.45
_____

18 £3.61
+ £2.27
_____

19 £4.25
+ £3.65
_____

4

Five children took part in the school sports day.

Each time they came first they were given 5 points.

Each time they came second they were given 3 points.

Each time they came third they were given 1 point.

20–24 Fill in the total number of points each child was given.

	1st	2nd	3rd	Total
Keith	✓✓	✓✓✓	✓	
Maria	✓✓✓	✓		
Martin	✓✓✓	✓	✓✓✓	
Tanya	✓✓	✓✓	✓✓	
Ian	✓	✓✓✓	✓✓✓	

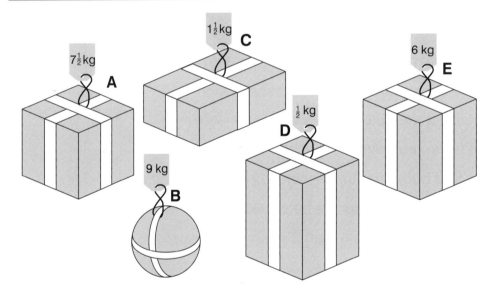

25 Which is the heaviest parcel? _____

26 Which is the lightest parcel? _____

27 How much heavier is B than A? _____

28 How much lighter is D than E? _____

29 How much heavier is E than C? _____

30 I save 50p each week.
How long will it take me to save £10.00? _____

Paper 22

1–6 Put the letters **a** to **f** in the correct place in the **Venn diagram**.

Shapes

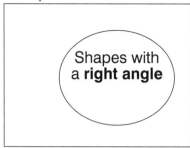 a b c

 d e f

7–8 Which shapes have an **odd number** of sides? _____ and _____

9–11 Which shapes have an **even number** of **vertices**?

_____ and _____ and _____ | II

12 A dinner lady puts 4 rows of pasties in the oven. There are 6 pasties on each of 4 rows.

How many pasties does she bake? _____ | I

This sign < means less than and this sign > means greater than.

Put one of these signs in each of the spaces below.

13 (7 + 2) _____ 8

14 (5 + 4) _____ 10

15 (2 + 6) _____ 7

16 (3 + 5) _____ 6 | 4

Here are some clocks for the times Paolo did things last Saturday.

Draw the hands of each clock, taking care to put them in the right place.

17 Paolo got up at 7:45 a.m.

18 He had breakfast at 8:30 a.m.

19 He went to football at 9:10 a.m.

20 He had lunch at 2:05 p.m.

21 Paolo went shopping with his mum at 2:40 p.m.

22 He ate again at 5:25 p.m.

23 He played on the computer at 6:20 p.m.

24 He went to bed at 8:35 p.m.

8

There are 25 children in the class. They made this **pictogram** to show how they are going to spend their holidays.

Seaside	👤	👤	👤	👤	👤	👤	👤	👤	👤	👤
Camp	👤	👤	👤							
Farm	👤									
Abroad	👤	👤	👤	👤	👤	👤	👤			
Touring										

👤 = 1 person

number of children

25 How many are going to the seaside? _____

26 How many are going to a farm? _____

27 How many children are going abroad? _____

28 How many are going to camp? _____

29 Fill in on the chart the number of children who are touring.

5

30 Write one thousand and seven in figures. _____

I

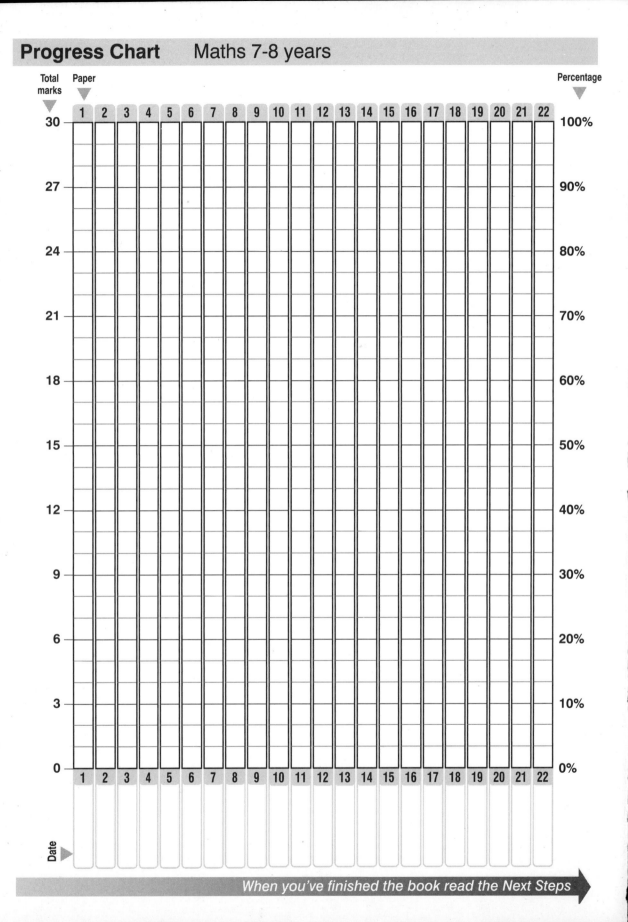